Eliminati.. ..

Electrical Noise

from 30kHz to 30MHz

Second edition by
Don Pinnock, G3HVA

There are two kinds of radio amateur, the small minority who deal with their noise problems and the vast majority who don't.

If this booklet converts just a small proportion of the vast majority, the effort made in compiling it will have been well worthwhile.

radio**today**

a Radio Today publication

Published by Radio Today as a trading name of the Radio Society of Great Britain, 3 Abbey Court, Fraser Road, Priory Business Park, Bedford MK44 3WH

First edition published 2010

This edition first published 2015

ISBN 9781-9101-9314-3

Publisher's notes

The opinions expressed in this book are those of the author and not necessarily those of the RSGB. While the information presented is believed to be correct, the author, publisher and their agents cannot accept responsibility for consequences arising from any inaccuracies or omissions.

Working on high voltage electricity supplies can be dangerous and should only be done by someone suitably qualified. Care must be taken to comply with the regulations relating to the mains power supply in your country.

Typography and design: Mike Dennison, G3XDV, Emdee Publishing
Cover design: Kevin Williams, M6CYB
Production: Mark Allgar, M1MPA

Printed in Great Britain by Latimer Trend of Plymouth

Contents

About the Author

Don Pinnock has always been interested in interference problems, and was a member of the very first RSGB TVI committee which was set up during the 1960s.

He has written the following articles which were published in *RadCom*: '*Where TVI is a problem, build this top band to ten SSB transmitter*' (Jan/Feb 1970) and '*A high-power 3-500 HF linear amplifier*' (May/June 1997).

Important Warnings:

- Working on high voltage electricity supplies can be dangerous and should only be done by someone suitably qualified. Care must be taken to comply with the regulations relating to the mains power supply in your country.

- In today's litigation-prone environment, it is RSGB advice that you do not modify other people's equipment. Working on high voltage electricity supplies can be dangerous and should only be done by someone suitably qualified.

Preface

*So-called background noise and band noise is frequently local noise
which the operator can't be bothered to deal with.*

The environment in which we live is now teeming with sources of electrical noise which are making it ever more difficult to pursue our hobby, particularly on the lower-frequency bands. This problem is a frequent topic of conversation on those bands, and one has only to listen for an hour or two to realise that many amateurs are suffering from long periods of high-level noise.

From time to time, comments have been heard to the effect that particular areas of the country are subject to periods of widespread noise. The scenarios depicted may even be of some comfort to the parties concerned, since the extensive nature of the problem would effectively absolve them from taking corrective action. However, such comments are almost invariably untrue, since noise from a single source is rarely heard over a wide area, rarely derives from the local airfield or radar station, and nearly always originates from sources which are much closer to home, probably no more than 100 metres away.

With their technical background, the majority of radio amateurs and listeners should be quite capable of tracking down noise sources, then dealing with the parties involved or, in the event of a negative response, reporting the matter to OFCOM. In spite of this, however, there are still many who avoid the issue completely by operating mobile, portable or even via the Internet!

It is hoped that this book will encourage many radio amateurs and listeners to improve their domestic scenario, and that it might even encourage members of radio clubs throughout the country to purchase the necessary equipment and help each other to overcome the problems involved. Perhaps the RSGB should award an annual certificate of merit to the radio club with the largest number of members operating from home!

Don Pinnock, G3HVA

Acknowledgements
I would like to dedicate this book to my old friend G3KAA, a top-notch DXer until his death in the year 2000, and to thank my friends G3MEA, G3MSL, G3MUI, G3TDH and G4BOH for their unstinting help and positive criticism during its preparation.

1

Introduction

Purpose of the Book

The main purpose of this document is to specify ways and means of eliminating the problem of electrical noise or interference emanating from man-made sources and their associated component parts.

The document covers electrical noise problems which are likely to be encountered on all bands from 10 to 160 metres.

Types of Interference or Noise

The terms 'QRM' and 'QRN' are often used indiscriminately on our bands by old-timers and newcomers alike. It is therefore essential that the various types of interference or noise which we are likely to encounter should be defined.

Broadband atmospheric interference, static or QRN

This is broad-band noise generated in the Earth's atmosphere, typically by lightning.

Co-channel and adjacent-channel interference or QRM

This is interference on a given channel which is caused by another station or stations on the same channel or an adjacent channel.

Electrical interference or noise

This is broad-band or narrow-band noise generated by man. It can be classified as follows:
- **Broad-band noise:** Noise, of the same or slightly varying amplitude, spreading across a band of frequencies. It either has no peaks at all in its spectrum, or broad peaks spaced at regular intervals. The spacing of the peaks indicates the fundamental frequency.
- **Narrow-band noise:** Noise occurring on one or more spot frequencies, and accompanied by little or no detectable noise between those frequencies.

Other Definitions Applicable to Noise

Noise source

Domestic or industrial equipment, including switch-mode power supplies, which generate any of the above-mentioned types of electrical noise.

Noise carrier

House wiring, domestic antenna systems, overhead power-distribution cables, sub-stations and overhead telephone lines which conduct electrical noise originating from noise sources.

Domestic antenna systems, overhead power-distribution cables and overhead telephone lines may radiate to a limited extent, depending on their configuration.

Noise-free or relatively noise-free zone

The part of a property which is noise-free or relatively noise-free. It will not necessarily be the end of the garden, but rather an area which is equidistant from, and at least 16m from, surrounding noise carriers.

Definitions Applicable to Antennas

Feed point

Whether a feeder consists of 75 or 300 ohm twin lead, 450 ohm ladder-line or 50 or 75 ohm coaxial cable, the feed point is the interface between:
- An antenna and its feeder,
- A driven element and its feeder in the case of a quad antenna, or
- The 34ft stub and feeder in the case of a G5RV antenna.

Inverted G5RV antenna

A G5RV antenna with an inverted-vee configuration.

Ladder-line

The name commonly used for 450 ohm balanced feeder.

Twin feeder

The name commonly used for 75 ohm and 300 ohm balanced feeder.

Units Used for Measurements

Noise levels

Since the noise levels specified in this book are not subject to large ranges of variation, but differ by relatively small amounts in each case, they can be expressed more conveniently in linear rather than logarithmic terms. The S-meter readings taken on the Icom IC-756 Pro 3 test transceiver, generally considered to be one of the quietest currently available in its price range, are therefore used, and the symbols '<' and '>' mean 'just below' or 'just above' the readings shown. The conversion rate for those who prefer to work in decibels is 6dB per S-point.

Distances and frequency bands

These are generally specified in metres.

Antenna heights and lengths

These are generally specified in feet.

Measuring and Monitoring the Local Noise Level

The fast AGC setting is by far the best option for measuring and monitoring the local noise level. Even in a band which is teeming with signals, the fast mode reduces the hang time and allows the S-meter to drop to the local noise floor. Noise measurements should quote bandwidth, AGC setting (fast) and preamplifier gain.

2

Sources and characteristics of noise

Sources of Broad-band Noise with Peaks

Battery chargers
Description of noise:
These devices generate very sharp spikes from Medium Wave to 20MHz. The spikes can be taken out using a noise blanker, but should preferably be suppressed in accordance with Chapter 9

Coverage:
Up to 30m or more.

Cure:
Install clip-on ferrites on interconnecting leads and ferrite rings on mains leads in accordance with Chapter 9.

Cable TV systems
Description of noise:
Street cubicles normally contain a switch-mode power supply which may well cause interference on the HF bands due to common-mode signals on the screen of the coaxial cable.

Coverage:
Up to 30m or more.

Cure:
Use a short lead to connect the screen to the mains earth or, better still, to an independent earth. This will reduce noise on 160, 80 and 40m considerably (typically from S7 to S2).

CD / DVD / video players and recorders
Description of noise:
These units may contain linear or switch-mode power supplies which run continuously, even when the units are switched off. See switch-mode PSUs.

Coverage:
Up to 30m or more.

Cure:
Install clip-on ferrites on interconnecting leads and ferrite rings on mains leads in accordance with Chapter 9.

Compact fluorescent lights

Description of noise:

These low-energy lights generally contain an 'electronic transformer' and generate more noise than large conventional fluorescent lights. See switch-mode power supplies.

Coverage:

Up to 30m or more.

Cure:

Replace 'electronic transformer' with a linear toroidal type.

Computers, peripherals and video games

Description of noise:

These units may contain linear or switch-mode power supplies which run continuously, even when the units are switched off. Some makes of computer are particularly poor in this respect. See switch-mode PSUs.

Coverage:

Up to 30m or more.

Cure:

Install clip-on ferrites on interconnecting leads and ferrite rings on mains leads in accordance with Chapter 9.

Fax machines

Description of noise:

These machines may well run 24 hours a day. See switch-mode PSUs.

Coverage:

Up to 30m or more.

Cure:

See switch-mode power supplies.

Halogen spotlights

Description of noise:

These low-voltage lights may be powered by 'electronic transformers' which generate noise. See switch-mode power supplies.

Coverage:

Up to 30m or more.

Cure:

Replace 'electronic transformers' with linear toroidal types.

In-house power-line telecommunication (commercial or domestic)

Description of noise:

In-house PLT (see Chapter 4) uses OFDM to modulate data onto a series of carriers in the frequency range 3-30MHz. When an in-house system transfers data, it radiates broad peaks with a sound like a frying noise. The sound is interrupted very briefly at random intervals.

Coverage:

Up to 30m or more.

Cure:

Use automatic notch function to eliminate carriers in the range 3-30MHz. Complain directly to OFCOM, in accordance with Chapter 11, and if it fails to resolve the problem, complain to the local electricity authority.

Office equipment in general

Description of noise:

This equipment may well run throughout office hours or 24 hours a day. See switch-mode power supplies.

Coverage:

Up to 30m or more.

Cure:

See switch-mode power supplies.

Switch-mode power supplies in general

Description of noise:

These generate a square-wave at a frequency of 20 to 120kHz, producing harmonics with broad peaks and 100Hz modulation. The peaks are spaced at multiples of the switching frequency, tend to drift with time, and may merge on the higher HF bands.

Coverage:

Up to 30m or more.

Cure:

Replace them with linear power supplies.

Television sets in general and associated items such as set-top boxes for cable TV, digital terrestrial TV, satellite TV, video on demand (ASDL) and mast-head amplifiers

Description of noise:

The power supply of some TV sets remains operative in the stand-by mode. The switching frequency of some models changes to around 1.75 - 2MHz in the stand-by mode. With set-top boxes, the switch-mode power supply runs continuously. See switch-mode power supplies.

Coverage:

Up to 30m or more.

Cure:

Install clip-on ferrites on interconnecting leads and ferrite rings on mains leads in accordance with Chapter 9. If these measures fail, complain directly to OFCOM, in accordance with Chapter 10, so that it can take action to enforce the non-interference condition of the TV licence.

Plasma screen television sets

Description of noise:

Plasma screen television sets may cause interference to amateur radio reception, mainly in the HF bands. The noise peaks are spaced some 60kHz apart and tend to have very sharp spikes.

Coverage:

Up to 30m or more.

Cure:

Install clip-on ferrites on interconnecting leads and ferrite rings on mains leads in accordance with Chapter 9. If these measures fail, complain directly to OFCOM, in accordance with Chapter 11, so that it can take action to enforce the non-interference condition of the TV licence.

Security lights

Description of noise:

Security lights may well contain a switch-mode power supply, and clip-on ferrites should thus be installed on the mains lead to each light as specified.

Coverage:

Up to 30m or more.

Cure:

Install clip-on ferrites on the mains lead to each light. The ferrites should be as close as possible to the light itself.

Touch lamps

Description of noise:

These lamps produce emissions similar to switch-mode power supplies, but with a fundamental frequency around 190kHz. Although they do not radiate noise as such, they are easily switched on and off by radio transmissions.

Coverage:

Up to 30m or more.

Cure:

Move affected lamps away from large metallic objects such as windows (double glazing), bed frames, power outlets, other power leads, etc. Ferrite rings and clip-on ferrites will rarely effect a cure. A slight reduction in power, say 25%, may effect a cure.

Sources of Broad-band Noise Without Peaks

Appliance switch contacts

Description of noise:

Arcing which occurs when contacts make or break may generate widespread noise.

Coverage:

Up to 100m or more.

Cure:

Clean the contacts and install appropriate capacitors to suppress any residual noise.

Dimmer switches

Description of noise:

On rare occasions, these switches produce noise similar to that generated by fluorescent lamps. The noise level is higher when the lamp concerned is dimmed.

Coverage:

Up to 30m or more.

Cure:

Install appropriate capacitors to suppress noise.

Electric blankets

Description of noise:

See thermostats.

Coverage:

Up to 100 m or more.

Cure:

See thermostats.

Electric fences

Description of noise:

These may produce a regular clicking noise, once or twice a second.

Coverage:

Up to 1km or more.

Cure:

On systems using a single bare conductor, the noise is due to flash-over at faulty insulators. The cure is to ensure that the landowner is aware of the fault and that the insulators are replaced. On systems using plastic rope or tape containing multiple strands of wire, arcing is caused by broken strands, although rain may reduce this noise temporarily. Breaks cannot be located using a receiver, but must be located directly by ear. The cure is to ensure that the landowner is aware of the fault and that the broken strands are repaired.

Electric motors

Description of noise:

These are used in dryers, drills, lawn mowers, toys, fans, food mixers, hair dryers, hedge cutters, model railways and sewing machines. They produce a rhythmic sound that varies with the motor speed.

Coverage:

Up to 30m or more.

Cure:

Install appropriate capacitors to suppress noise.

Engine-tuning systems

Description of noise:

These may generate widespread noise in the vicinity of garages.

Coverage:

Up to several hundred metres.

Cure:

Ensure that the garage owner is made aware of the noise and that he takes action to suppress it.

Flashover on electrical power distribution systems

Description of noise:

The resulting noise may be very widespread.

Coverage:

Up to 1km or more.

Cure:

Track down the source and notify the local electricity authority.

Fluorescent lights

Description of noise:

Large conventional types may generate noise modulated with a 100Hz buzz, usually on the lower HF bands. A worn-out tube flickering at 50Hz may increase the level of interference.

Coverage:

Up to 30m or more.

Cure:

Replace worn-out tubes and install appropriate capacitors to suppress any residual noise.

Heating control systems

Description of noise:

See thermostats.

Coverage:

Up to 100m or more.

Cure:

See thermostats.

Heating pads

Description of noise:

See thermostats.

Coverage:

Up to 100m or more.

Cure:

See thermostats.

Ignition systems in general

Description of noise:

These systems produce a rhythmic sound (buzz) that varies with the motor speed.

Coverage:

Up to 30m or more.

Cure:

Clean contacts and install appropriate capacitors to suppress any residual noise.

Internet / telephone connections

Description of noise:

Noise originating from the computer itself or from associated modems or routers be radiated via the telephone line.

Coverage:

Up to 30m or more.

Cure:

Install ferrite rings or clip-on ferrites on the telephone line.

Neon signs

Description of noise:

These signs may be a prolific source of noise due to the following: HV sparking to nearby metal objects, loose electrical connections or flickering caused by loss of gas pressure. The noise may be very widespread.

Coverage:

Up to 1km or more.

Cure:

Ensure that the owner is made aware of the noise and that he takes action to suppress it.

High-voltage power lines (National Grid)

Description of noise:

These may generate noise due to corona discharge from the cables. The noise may be greatly increased by arcing at faulty insulators, lightning arresters and transformers, and loose or unbonded hardware. In such cases, it may be very widespread.

Coverage:

Up to 1km or more.

Cure:

Track down the source and notify the local electricity authority.

Petrol engines

Description of noise:

Any noise will generally be caused by the ignition system.

Coverage:

Up to 30m or more.

Cure:

See ignition systems in general.

Thermostats in general

Description of noise:

These may arc for 1 to 30 seconds and produce noise with 100Hz modulation. Arcing may occur at 5-20 minute intervals, or 2-3 times a minute in some cases. The most common source, especially in winter, is a central heating boiler or room thermostat. If the noise is heard 24 hours a day throughout the year, a refrigerator or freezer thermostat may be the source.

Coverage:

Up to 100m or more.

Cure:

Clean the contacts and install appropriate capacitors to suppress any residual noise.

Vehicle ignition systems

Description of noise:

In shacks located near a busy road, ignition noise may cause problems with weak-signal reception.

Coverage:

Up to 30m or more.

Cure:

Ensure that the shack is as far away from the road as possible. If the noise persists, move house!

Water conditioners

Description of noise:

One type generates approximately 164 regular modulated beats per minute across the lower HF bands. Another type generates a sequence of audio-frequency tones with harmonics up to 28MHz in some cases.

Coverage:

Up to 30m or more.

Cure:

Move antennas away from the noise source.

Sources of Narrow-band Noise on a Number of Frequencies

Alarm systems in general

Description of noise:

These systems produce harmonics, with slight modulation, which tend to drift.

Coverage:

Up to 30m or more.

Television receivers in general

Description of noise:

Television receivers produce time-base harmonics spaced at 15.625kHz intervals, and the sound changes with the picture content. The interference is usually very local.

Coverage:

Up to 30m or more.

Older television receivers with satellite receiver and dish

Description of noise:

Some of these systems produce strong spurious signals on 3914kHz (S8), 3836kHz (S6), 3774kHz (S6), 3696kHz (S8), 3557kHz (S6), 3479kHz (S9), 3401kHz (S7), 3340kHz (S6), 3262kHz (S8), etc.

Coverage:

Up to 30m or more.

Sources of Narrow-band Noise on One Spot Frequency

PC crystal oscillator carriers

Description of noise:
These carriers are to be found on or around 14318kHz and are usually very weak and local.

Coverage:
Up to 30m or more.

Area Affected by Electrical Interference or Noise

The noise coverage shown in the above paragraphs is only very approximate, since it depends not only on the degree of suppression included in the equipment (see Chapter 4), but also several other factors. The following are particularly important:

Energy Involved at Noise Source

HV power lines (National Grid)

High-voltage power lines are generally accompanied by a wide swath of noise extending up to half a kilometre or more on either side. Faults which occur on these lines often involve a high energy level, and the resulting noise may well be distributed over a wide area, particularly if it is radiated as well as conducted.

High-current domestic devices

Faulty domestic thermostats, electrical connections and switches may well cause interference up to 100m or more.

Means by which Noise is Distributed (Noise Carriers)

Under normal circumstances, an amateur radio antenna is likely to pick up noise from various noise carriers, that is to say:

House wiring

Noise generated by actual noise sources (domestic equipment and switch-mode power supplies) is generally conducted throughout a property via the mains wiring, including the earth lead. In such cases, the entire house wiring acts as a noise carrier. The same applies to the house wiring of neighbouring properties.

Domestic antenna systems

Noise deriving from actual noise sources (domestic equipment and switch-mode power supplies), which is normally conducted via the mains supply, may well cause interference up to 30m or more if it is radiated via an antenna system as well. In such cases, the entire house wiring will act as a noise carrier and the antenna system as both carrier and radiator, thus exacerbating the noise problem.

Overhead power-distribution cables, sub-stations and overhead telephone lines

Overhead power-distribution cables, sub-stations and overhead telephone lines are not inherent noise sources, and only generate noise themselves if they are faulty. Thus, any noise they radiate will nearly always originate from an actual noise source, domestic or industrial, in the vicinity. In such cases, the overhead systems or sub-stations will act as noise carriers.

Electric fences

Faults on electric fences do not involve a high energy level, but the resulting noise may well be conducted and radiated over a wide area because of the length of the conductors involved.

Characteristics of Received Noise

The noise heard will differ in accordance with the source.

Arcing or flash-over on HV power lines, power-distribution systems or faulty switches and thermostats

This noise will be heard as a distinct crackle in the AM, CW and SSB modes.

Switch-mode power supplies

This noise will be heard as a distinct hum in the AM mode, especially at peaks.

Unsuppressed ignition systems, motors, etc.

This noise will be heard as a rhythmic sound (buzz) which varies with the speed of operation.

Power-line telecommunication (PLT)

This noise will be heard, 24 hours a day, as a continuous modulated carrier over large parts of the short-wave frequency spectrum. The amateur and broadcast bands are supposed to be protected by notches.

When data is being transferred, it will result in broad peaks with a frying sound in the SSB and CW modes. If the noise carrier is nearby, the peaks may well be heard throughout the short-wave frequency spectrum, and will be interrupted very briefly at random intervals.

Main Factors Determining the Received Noise Level

Spacing between the amateur radio antenna and noise carriers

The main factor determining the received noise level will almost certainly be the horizontal and vertical spacing between an amateur radio antenna and any of the above-mentioned noise carriers. In fact, it can safely be assumed that the inverse-square law applies, and that the level of received noise decreases in proportion to the square of the distance from those noise carriers.

Height of noise carriers

Noise carriers can have a worse effect, in particular on the HF bands, when located upstairs.

3

Choosing a property & setting up a station

Property Scenarios

Optimum location

A property which is so isolated that it is unaffected by surrounding noise carriers.

Good location

A property where a garden shed or caravan can be installed in a relatively noise-free zone, and where noise from surrounding carriers can be eliminated, if necessary, using a noise-cancelling device and/or loop antenna (preferably both).

Average location

A property where a garden shed or caravan cannot be installed in a relatively noise-free zone, but where noise from surrounding carriers can nevertheless be eliminated, if necessary, using a noise-cancelling device and/or loop antenna (preferably both).

Poor location

A property where a garden shed or caravan cannot be installed in a relatively noise-free zone, and where the quantity and proximity of surrounding noise carriers makes remedial action almost impossible.

Choosing a Property

Most of us are restricted by our budgets when purchasing a property, and newly licensed amateurs in particular are often unable to purchase one which is ideal for amateur radio purposes. For those who are fortunate, *a carefully chosen location can make all the difference to the ambient noise*, and the following guidelines in particular should be followed when purchasing:

A property should never be purchased in:
- the vicinity of high-voltage overhead power lines, that is to say in viewing range (see Chapter 2)
- a road or street with overhead power-distribution cables
- a road or street with overhead telephone lines
- a road or street with a steep slope, since reception could be spoiled by a high level of ignition noise (traffic lights and similar obstacles exacerbate the problem)

Nor should a property be purchased in:
- the vicinity of factories, garages, industrial estates, public houses and similar noise sources
- the close vicinity of another radio amateur or a low-frequency (LW or MW) radio station

Before a property is purchased, the local electrical environment should always be checked using a portable all-band receiver, particularly in the vicinity of noise carriers such as street lamps, sub-stations, etc.

Type of Property

Since antenna height can affect the level of unwanted noise, it is important that the following be borne in mind when a property is purchased.

Bungalows

Apart from the fact that bungalows are generally less cost-effective than houses in terms of usable space and floor area, they have the following disadvantages:
- They are generally unsuitable for mounting poles to support antennas at heights exceeding 25ft
- Their ability to hide antennas, poles, towers, etc located in the back garden is minimal

Houses

Houses are not only more cost-effective than bungalows, but they have the following advantages:
- They are generally suitable for mounting poles to support antennas at heights up to 40ft or so
- They are able to hide antennas, poles, towers, etc located in the back garden

Type of roof

The suitability of a house or bungalow for mounting poles to support antennas at a given height will depend to a large extent on the type of roof involved.

Gable roof: This has just two slopes at more than 20 degrees to the horizontal and is ideal, since the gable walls at either end can be used to mount a pole supporting antennas well above apex height.

Hipped and similar roofs: This type of roof slopes in all directions and cannot therefore be used to mount a pole supporting antennas well above apex height.

Use of chimneys for antenna supports

The disadvantage of hipped and similar roofs, and mid-terrace houses with no gable walls, can be overcome to some extent by mounting antenna poles on chimneys. However, if the antenna system needs to be altered or repaired, access could be difficult or the services of a specialist antenna erector could be required, especially during inclement weather.

A gable roof is ideal for mounting a pole

A hipped roof presents some difficulties

Setting up a Station at Home

Key factors to be taken into account

In view of the environment in which we live, it is essential that consideration be given to the following when setting up a station:
- The position of the shack itself
- The type of antennas and feeders used, and their vertical and horizontal clearance from noise carriers
- The method used to filter the mains supply to the house and shack

Disadvantages of operation from the house itself (noise zone)

Even if a property is ideal from the propagation viewpoint, ie located on top of a hill, by the sea, in open country, etc., operation from the house itself may still involve a number of problems and restrict one's operation.

Ground floor operation: This is not recommended because of the level of electrical noise emanating from television sets, fax machines, computers, etc. This problem has become worse during the last decade due to the widespread use of switch-mode power supplies, not only for equipment but also for lighting.

Upstairs operation: This is equally difficult and may even be worse because of the problems involved in providing a short earth lead. If upstairs operation is unavoidable, coaxial cable should be used for the earth lead. Both the inner and outer conductors should be connected to ground via a copper earth rod, and the inner conductor alone connected to the equipment. The actual earth lead (the inner conductor) will thus be shielded.

If any of the above-mentioned problems occur, it is well worth considering the installation of a good quality garden shed or caravan.

Advantages of operation from a garden shed or caravan in a relatively noise-free zone

Provided that every precaution is taken to maximise security and minimise damp and other adverse effects, operation from a garden shed or caravan is a much better proposition because of the electrical and physical isolation which it provides.

Position of shed or caravan: The shed or caravan should be located:
- in an optimum position for antenna feeders and/or, preferably,
- in a noise-free or relatively noise-free zone equidistant from noise carriers

Power and other cables: Power and other cables to the shed or caravan may:
- be suspended above ground
- run along the ground or, preferably,
- run underground

The cables should be filtered at each end to eliminate conducted noise.

Earth lead: The earth lead should be as short as possible.

Domestic and safety aspects: Family life will be unaffected by the operation of the equipment, even at night, and the latter will present minimal danger to family, friends, animals, etc.

Warning:

Working on high voltage electricity supplies can be dangerous and should only be done by someone suitably qualified. Care must be taken to comply with the regulations relating to the mains power supply in your country.

Setting up a Station Away From Home

Existing situation on the lower bands

Many portable and special-event stations working on the lower frequency bands suffer from very high noise levels on receive, and the operators spend long periods of time asking for repeats or doubling with their opposite numbers. It is obvious to any seasoned listener that the people setting up such stations, generally club members, have probably never operated from home and thus have no idea how to avoid or resolve electrical noise problems.

Failure to monitor selected buildings or sites beforehand

Since the people setting up temporary stations are often unaware of potential noise problems, they concentrate most of their efforts in erecting antennas as high as possible above the building or site from which they are going to operate, and hope for the best. Unfortunately, public buildings and sites in particular often contain a large number of noise sources and the building or site in which the station is set up may well turn out to be a virulent noise carrier.

Actions which can be taken beforehand

Because noise problems cannot usually be cured by a quick fix on the day or days of operation, any group intending to set up a portable or special-event station should check and monitor the chosen building or site *beforehand*. A small portable receiver such as the Sony ICF SW7600G can be used to check the noise level throughout the short-wave spectrum. If the noise level is low, a mains filter such as the Belling-Lee type SF4240-16/01 may well reduce it further. However, if the amateur bands are subject to a high level of electrical noise, every effort must be made to find a noise-free zone outside the building or site. It is better that operation should take place from a caravan or tent if the clarity of reception "sells" our hobby to the general public.

4

Dealing with electrical noise

Locating Noise Sources and Eliminating Noise

If, in spite of all the precautions taken when setting up a station, the level of electrical noise on receive is unacceptable, serious consideration must then be given to procedures for locating the noise sources and eliminating the noise.

In the first instance, a simple sensing device, consisting of a 5-10ft length of thin coaxial cable terminated with a small single-turn or multi-turn loop, can be connected to the external antenna socket of a small all-band receiver, such as the Sony ICF SW7600G, with a view to detecting noise sources in one's own and neighbouring properties.

The problems caused by electrical noise may appear to be daunting, but if the basic guidelines shown in this document are adhered to, and the direction-finding and tracking operations referred to in Chapters 6 and 7 are carried out as specified, the received noise can be reduced to a very low level indeed (S1 or lower).

Power-line Telecommunication (PLT)

There are two types of PLT:

Access PLT (not implemented in most of Europe)

With access PLT, the service provider uses low-power radio signals to transmit broad-band data to commercial or domestic premises via the National Grid and power distribution systems. This type of PLT would almost certainly cause widespread interference, especially in areas where overhead power distribution cables are used. Fortunately, it is unlikely to be authorised in the foreseeable future.

In-house PLT

With in-house PLT, the service provider uses OFDM type multiplexing to modulate data onto a series of carriers in the frequency range 3-300MHz. It transmits the broad-band data, via the local telephone line, to commercial or domestic premises where it is transferred via a modem to the mains wiring. This provides the following facilities in the properties concerned:

- The reception and transmission of broad-band data via any power outlet
- In-house networking between devices connected to power outlets.

Although the system can be programmed to cover only certain parts of the short-wave spectrum, using notches to protect the amateur or broadcast bands, it is extremely detrimental to radio amateurs and short-wave listeners for the reasons specified below.

Disadvantages of in-house PLT

Whereas most man-made interference to radio services has traditionally come from unintentional sources that generate RF noise as a by-product of their operation, in-house PLT causes widespread interference over a large part of the short-wave frequency spectrum for the following reasons:

- It generates signals intentionally throughout that spectrum, for data transmission and networking purposes, and uses in-house mains wiring as the carrier.
- Since in-house mains wiring was never designed to carry high-frequency signals and is rarely balanced (the live wire often runs alone to and from light switches for example), the system acts as a high-level noise carrier.
- Unwanted noise generated in a property with PLT will almost certainly be conducted via overhead or underground power distribution cables to neighbouring properties.
- The notches included to protect parts of the short-wave spectrum, such as amateur and broadcast bands, may lead to intermodulation under certain conditions.
- The notches will always be susceptible to change due to pressure from broad-band users. The more notches that are included in the system, the narrower the overall bandwidth and the lower the bit rate. In the long term, this will be unacceptable to broad-band users who will always require higher bit rates, one of the main attractions of broad-band Internet access.

To make matters worse, an in-house PLT system may be operative 24 hours a day.

Actions which could alleviate the problem of in-house PLT

Several actions can be taken with a view to alleviating the problem, namely:

In the house: The inclusion of a mains filter or blocking capacitors at the point where the mains supply enters the property (see Chapter 6), particularly if the victim's property and the noise source are on the same phase of a three-phase supply.

In the shack: The use of the ANF (automatic notch function) on transceivers or receivers which include this facility. The optional UT-106 unit in the Icom R75E receiver is very effective in eliminating carriers.

Complaints: A radio amateur is responsible for tracking down a noise source before submitting a complaint to OFCOM in accordance with Chapter 11. If OFCOM fails to solve the problem, he should then submit a complaint to the local electricity authority. In the author's experience, these authorities take a very dim view of anyone contaminating the mains supply, whether intentionally as in the case of PLT, or otherwise.

Other action: Failing all else, the insertion of articles in local magazines or newspapers declaring the use of PLT to be a potential hazard on the grounds that continuous close-range exposure to an electronic smog covering a large part of the radio spectrum could be detrimental to health. The articles should preferably be submitted by someone other than a radio enthusiast!

Developments which could eventually alleviate the problem of in-house PLT

Desktop computers are gradually becoming less popular. Many people now want laptop computers because they are smaller and portable. Laptops are ready fitted with a wireless connection which isn't a problem. As a result, the above-mentioned power-line devices will probably have only a limited take-up and will be abandoned as people demand portability.

Switch-mode Power Supplies

Advantages

Switch-mode power supplies are now in common use throughout the community. It is accepted that they are more efficient than linear supplies, and are usually lighter and more compact. A well-designed unit, one which is thoroughly screened and adequately filtered at both input and output, will rarely produce unwanted noise or interference.

Disadvantages

Unfortunately, in order to reduce cost, many switch-mode supplies are designed and manufactured with little regard for RF interference. In fact, the vast majority of the units currently sold as 'electronic transformers' on the domestic market (for instance, at D-i-Y stores) are nothing less than broad-band noise generators. The same can be said of many domestic appliances containing or using this type of supply. Many radio amateurs are therefore plagued by unwanted electrical noise, particularly on the HF bands.

Dealing with switch-mode supplies on your own property

It is easy to deal with electrical noise generated on your own property. Just replace the offending supplies with linear supplies of the same rating.

Dealing with switch-mode supplies outside your property

As with in-house PLT, the radio amateur is responsible for tracking down a noise source before submitting a complaint to OFCOM.

Individual switch-mode supplies: If you suffer from electrical noise generated by switch-mode power supplies on a neighbour's property, and you are absolutely sure of the source, you should offer to replace them with linear supplies, of the same rating, *free of charge*. OFCOM should only be contacted if the neighbour concerned is unduly difficult to deal with.

Multiple switch-mode power supplies used for down-lighting: If you suffer from electrical noise generated by multiple switch-mode supplies (the so-called 'electronic transformers') on a neighbours premises, in a bathroom or kitchen for example, and you are absolutely sure of the source, complain directly to OFCOM and *offer to pay the additional cost involved in installing linear supplies*. If OFCOM fails to rectify the problem for any reason, complain to the local electricity authority.

Warning: It is important that the supplies be replaced by a competent registered electrician who should check the overall condition of the system and carry out any necessary corrective work before making changes.

If the problem is not resolved, it is inevitable that all houses on the contaminated phase of the electricity supply will remain badly affected by electrical noise, and will act as noise carriers to the detriment of amateur radio reception in the vicinity.

Power Supplies Available on the Amateur Radio Market

The amateur radio market itself is not free from inadequately suppressed examples of switch-mode technology, and some of the units advertised even include an offset control to permit the noise which they generate to be moved from the receive frequency. The use of such equipment can hardly be called technical progress!

Two linear power supplies suitable for amateur radio use

Switch-mode supplies are slightly cheaper than linear supplies but, while a 30A regulated linear DC supply is larger and heavier, it provides genuine noise-free operation for only £20 to £30 more. Surely, this is the only sensible choice for any shack, even for those who operate within the limited space of a cupboard or garden shed.

In fact, apart from those who are unable to operate from their home QTH, for whatever reason, and wish to work portable or participate in outside events where portability is a prime consideration, why would anyone want to purchase cheap noise-generating equipment?

Power supplies available on the domestic market

A limited market survey has shown that it is becoming ever more difficult to obtain power supplies which do not involve some form of switching action and hence RF interference.

One of the largest distributors of domestic electronic equipment, for example, markets two types of power adaptor, one "switch-mode" and the other "switching regulator". Neither type is noise-free and neither type is suitable for amateur radio purposes. Following a telephone enquiry, the company actually stated in writing that they could only supply one noise-free power adaptor which, according to their catalogue, is a switch-mode type!

Andrew Ikin of Wellbrook Communications recommends using noise-free power supplies. The catalogue of CPC (http://cpc.farnell.com/) includes a range of regulated linear power adaptors with various output voltages and at various currents.

Power supplies available on the Internet

One of my neighbours accidentally damaged the Dell 19V switch-mode supply used to power his Dell laptop computer, and immediately purchased a replacement of unknown make on the Internet. The label on the new unit (type N50/N17208) clearly stated that it was made in China and complied with Canada ICES-003/NMB-003.

Unfortunately, the replacement generated wide S9+ noise throughout the short-wave spectrum. The pulses could be heard up to 36MHz and at distances of up to 300 metres from the source. As soon as the neighbour was notified of the noise problem, he purchased the proper Dell replacement for his laptop and handed over the defective unit for destruction.

At present, little is done to prevent such defective equipment being sold in this country, whether in shops or on the Internet, and purchasers may well be enticed into buying such sub-standard rubbish. It is therefore recommended that power supplies should not be purchased on the Internet unless the supplier is known, reputable and fully aware of potential noise problems.

Equipment with Inadequate Noise Suppression

From time to time, electrical and electronic equipment is sold with inadequate noise suppression and, in some cases, with no noise suppression components at all. Such equipment will almost certainly fail to meet the minimum EC requirements in this respect.

Dealing with inadequately suppressed equipment on your property

It is easy to deal with inadequately suppressed equipment on your own property. Just return it to the vendor and demand a refund.

Dealing with inadequately suppressed equipment outside your property

If a neighbour is in possession of inadequately suppressed equipment, he should be persuaded, in his own interest, to return it to the vendor for a refund. If he fails to do so, a complaint to OFCOM may unfortunately be the only solution.

Badly Contaminated Phase of the Electricity Supply

Where a phase of the electricity supply is badly contaminated, a nearby radio amateur could be forgiven for thinking that he is surrounded by noise sources, since the noise will appear to come from several directions. However, while it is possible for several noise sources to affect a given area, experience over the last few years has shown that widespread interference tends to originate from just one or two virulent sources.

An example of a virulent noise source is a bungalow nearly 500ft away from the author's shack on a bearing of 50 degrees. The kitchen and bathroom contain a total of 21 down-lights, each with its individual switch-mode supply. The total noise generated in the bungalow is such that even the metal framework of a desk lamp radiates throughout the short-wave spectrum! Using the station quad antenna, wide noise pulses from the offending property can be heard on all bands from 14 to 28MHz.

The noise generated in the bungalow is conducted underground to other properties via the contaminated phase of the mains supply. The properties concerned are positioned on an arc extending from the offending bungalow to a house on a bearing of 250 degrees from the author's shack. Each property acts as a noise carrier and radiates the unwanted noise to some degree. The noise peaks in the house of a close neighbour on a bearing of 330 degrees. He allowed me to install a plug-in filter in his garage and the noise was almost entirely eliminated.

Dealing with a badly contaminated phase of the electricity supply

As shown above, electrical noise on a contaminated phase can be reduced considerably by inserting a plug-in filter into the contaminated phase at any suitable point, *not necessarily on the premises containing the noise source*.

The filter consists of a small diecast alloy box 5cm x 5cm x 3cm (available from Maplin), three mains-rated capacitors (470n between Line and Neutral, 33n between Line and Earth and 33n between Neutral and Earth), two small

Detail of the mains noise reducing filter. (NOTE: Construction of equipment designed for connection to the mains should not be carried out by someone who is inexperienced)

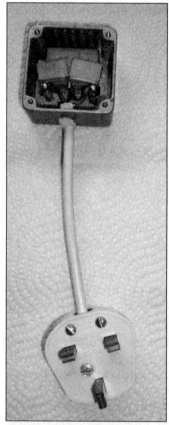

The completed mains filter

The filter in operation

> **Warning:**
> Working on high voltage electricity supplies can be dangerous and should only be done by someone suitably qualified. Care must be taken to comply with the regulations relating to the mains power supply in your country.

stand-off insulators, a 6" mains lead and a standard three-pin mains plug with a 3A fuse. The filter is pictured opposite.

Ancillary Noise Locating Equipment

1. Ferrite rod antenna mounted on a rotating disc. It can be turned quickly to ascertain which direction any electrical noise is coming from. It tunes from the top of the Medium Wave to around 7 MHz.

2. Loop/Ferrite Rod switch permitting selection of the following:
   - Wellbrook loop (with associated gain control)
   - MW loop
   - Ferrite rod antenna (as described above)
   - 144 MHz antenna

(below left) Rotating ferrite rod antenna

(below right) Loop / ferrite rod switch and noise antenna switches

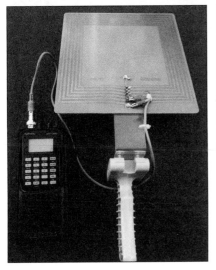

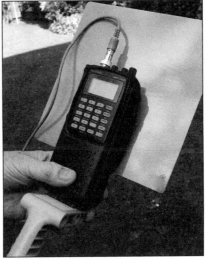

(left) Two views of of a Mizuho UZ77 printed-circuit loop antenna (no longer manufactured) used in conjunction with a Yupiteru MVT-7100 scanning receiver. This relatively unobtrusive combination can be used to check for noise emanating from properties in the vicinity

3. A pair of noise antenna switches permitting the following:
 - All noise antennas to be disconnected (OC position)
 - Any one or two noise antennas to be selected

Use of transceiver and receiver settings to reduce incoming electrical noise

The following actions usually result in incoming electrical noise being reduced to a reasonable level:

1. Switching the preamplifier out
2. Switching the attenuator in
3. Switching the notch filter in
4. Using the noise blanker
5. Using the bandwidth filter.

Unfortunately, although a very useful device for monitoring broad-band conditions and trends, the spectrum scope will not permit any direct reduction of noise whatsoever!

Noise immunity of antennas

Since our environment is teeming with sources of electrical noise, the noise immunity of an antenna system is of crucial importance. If a station is affected by local electrical noise, the antenna system should be selected and positioned bearing in mind the test results and conclusions of this chapter.

Actions Which Could Marginally Improve the Noise Immunity of Existing Antennas

End-fed and Windom antennas

Both end-fed and Windom antennas have low noise immunity, and the only way to reduce unwanted noise is to use a loop antenna for reception (see Chapter 7), or to replace them with a centre-fed type. Unfortunately, the use of a noise canceller may well introduce further problems (see Chapter 8).

Centre-fed antennas in general

With centre-fed antennas, unwanted noise can be reduced and reception improved by using a loop antenna and/or noise canceller, preferably both (see Chapters 7 and 8), or by making minor changes to the antenna itself. As shown in this chapter, such changes may or may not improve the antenna noise immunity, but should in any case be tried.

- **Inclusion of a balun at the feed point:** When any type of balanced feeder, whether 75 or 300 ohm twin feeder or 450 ohm ladder-line, is connected to coaxial cable, a balun should preferably be included to ensure that the antenna is balanced and prevent RF currents on the outer braid of the cable. A 1:1 balun should be used with 75 ohm twin feeder, and a 4:1 balun with 450 ohm ladder-line. Choke and sleeve type baluns, such as those shown on the next page, are recommended. The effectiveness of a balun can be verified by checking that the SWR does not change when the length of the coaxial cable is varied. Whether a balun is used or not, the 'hot' end of the antenna should be furthest away from any noise carriers.
- **Inclusion of a ferrite sleeve on coaxial cable at the feed point**, and reversal of the feed-point connection so that the house end of the antenna is connected to the inner conductor of the coaxial cable at the feed point.
- **Use of 75 ohm coaxial cable from the antenna itself** instead of 75 ohm twin feeder. A balun should obviously be included at the feed point.
- **Reversal of the feed-point connection** so that the house end of the antenna is connected to the braid of the coaxial cable at the feed point.
- **Earthing the braid of coaxial cable at the feed point.**

Antenna noise immunity may be improved by using a choke-type balun, such as any of those shown here. A sleeve balun, shown in (d) is recommended by the author
[Source: Radio Communication Handbook (RSGB)]

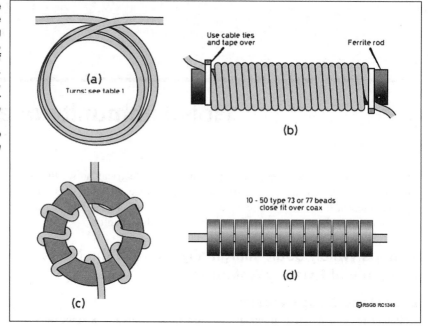

- **Earthing either side of twin feeder or ladder-line at the feed point.**
- **Positioning twin feeder or ladder-line** at an angle of 45 degrees to the ground, parallel to the ground and vertically.
- **Inclusion of clip-on ferrites on coaxial down-leads in the shack itself:** It is advisable to include clip-on ferrites on coaxial down-leads between the point where they enter the shack and the transceiver.

Vertical antennas

The following actions may be taken with a view to improving the noise immunity of vertical antennas.

- **Positioning the antenna:** The antenna should be positioned as far as possible from the nearest noise carrier.
- **Earthing the braid of coaxial cable** at the feed point.

Noise Immunity Tests

Arrangement of test site

As shown in the drawing opposite, the antennas under test ran from the top of a pole on the side of the house, a relatively noise-free zone, to a point on the tower near the end of the garden, a noise-free zone. This arrangement ensured that, as far as possible, none of the measurements was affected by noise from extraneous sources to the detriment of any particular antenna.

Equipment used for tests

The equipment consisted of the following:

- Icom IC-756 Pro3 transceiver with internal antenna tuner
- An SWR bridge. (If it is a 50 ohm bridge, use a 50-75 ohm Z-match to match a 75 ohm antenna. Alternatively use a 75 ohm bridge, eg Osker)
- A 75 ohm SWR/power meter in the antenna feeder.

Station antennas used for comparison purposes

The station antennas listed below were matched using the 50 ohm SWR bridge and 50-75 ohm Z-match with the transceiver internal tuner in or out of circuit as necessary.

Full-size inverted-vee dipole for 80m: This antenna was at 40ft, with 75 ohm twin feeder to a balun mounted on the tower, and 75 ohm coaxial cable to the shack. It has a high degree of noise immunity because the feeder and the ends are positioned for minimum noise pick-up. A further advantage is that it only requires a single support. The SWR at the centre of the band can be brought to 1:1 by increasing or reducing the length of the antenna at points near ground level.

Short inverted-vee dipole for 80m (with 40m traps): This antenna was at 38ft, with 75 ohm twin feeder to a balun mounted on the tower, and 75 ohm coaxial cable to the shack. It has a high degree of noise immunity because the feeder and the ends are positioned for minimum noise pick-up. A further advantage is that it only requires a single support. It is excellent for DX on 40m, and provides a bandwidth of 100kHz on 80m. The SWR at the centre of the 40m band and the chosen 80m segment can be brought to 1:1 by increasing or reducing the length of the 40m and 80m sections at points near ground level.

Two-element quad antenna for 20, 17, 15, 12 and 10m: An antenna with individually-switched driven elements and boom at 42ft.

Arrangement of the test site. The short inverted-vee for 80/40m is directly below the inverted-vee for 80m

Station antenna switching system

Antenna selector: The above-mentioned antennas are selected remotely using a *primary switch-box* mounted at the top of the tower.

Quad element selector: The driven elements of the quad antenna are selected remotely using a *secondary switch-box* mounted on the feed-point insulator of the quad itself.

Ledex switches: Both the primary and secondary switch-boxes contain a remote-controlled Ledex switch which not only selects the required antenna or driven element, but also shorts out all unused antennas and driven elements at their feed points. The switch-boxes have been airborne for some 43 and 30 years respectively and have fully complied with my requirement for absolute reliability! (For some strange reason, the amateur radio market has never taken advantage of the benefits offered by Ledex switches. Even the Japanese seem to have missed out!)

The Ledex switches are controlled by the unit shown overleaf. The front panel controls are as follows:

(left)   Detail   of the Ledex switch assembly mounted on Perspex and . . (right) . . the switch control unit

1. Top centre: Band-Selection switch. When this switch is set to the left-hand side, the momentary push-button (below left) can be used to select the 40m, 80m and 20m bands in that order. Further depression of the momentary switch will cause a return to the original setting. If the 20m band is selected, the band-selection switch can then be set to the right-hand side and the momentary push-button used to select the 17m, 15m, 12m and 10m bands in that order. Further depression of the momentary switch will cause a return to the original setting, the 20m band.

2. Middle row from left to right: Momentary push-button, on/off indicator lamp and mains on/off switch.

Feeders: All antennas and driven elements are connected via 75 ohm twin feeder. The input terminals of the secondary (quad) switch-box are connected via 75 ohm twin feeder to the 'quad' terminals of the primary (antenna) switch-box. The input terminals of the primary (antenna) switch-box are connected via 75 ohm twin feeder to a sleeve balun mounted just below it.

Sleeve balun: This balun is used to maintain the balance of the antenna system in use, and to prevent RF currents on the outer braiding of the 75 ohm coaxial cable between the balun and the shack.

Typical noise levels achieved using the station antennas

Broad-band noise: With the Icom IC-756 Pro3 transceiver in the SSB mode and the pre-amp off, the following are typical S-meter readings around midday.

80m:	S1.5
40m:	>S1
30m:	S0.5
20-10m:	S0

The highest readings (<S3) occur on 80m during cold damp evenings between October and March, presumably because the use of domestic equipment is at its peak.

Narrow-band noise: Narrow-band noise levels on 30, 40 and 80m reach anything between S1 and S6-7 at times, but are of no consequence since the affected channels do not have to be used and, in any case, the noise can easily be eliminated or reduced using the station Wellbrook loop antenna or MFJ noise canceller (see later in this book). The latter is easier to use with the internal tuner in circuit.

Test procedure

The IC-756 Pro3 transceiver has two antenna terminals on the rear panel and a single "Antenna" button on the front panel. This permits instant switching from Antenna 1 (station antenna) to Antenna 2 (antenna under test). The received noise level obtaining with a given station antenna and an antenna under test can thus be compared instantaneously on any frequency.

All the tests were carried out with the transceiver in the SSB mode and the pre-amp off. To prevent interaction, the antennas under test were installed one at a time. The tests took place on weekdays around midday to reduce the possibility of co-channel or adjacent-channel signals. Several sets of comparative S-meter readings were taken for each antenna under test. They were spaced a few days apart in order to take account of variations in noise level from the various local noise carriers. Although the readings varied somewhat over a given band, they all related to broad-band rather than narrow-band noise.

Initial antenna tests and conclusions

As stated above, all antennas under test ran from the top of a pole at least 15ft above the apex of the house (a relatively noise-free zone) to a point on the tower near the shack at the end of the garden (a relatively noise-free zone).

Antenna with a single-wire feeder 1:
End-fed (random length at 36ft with down-lead to shack)

End-fed (voltage-fed) antennas are probably the easiest to install (a lazy man's delight). Although they may radiate well, they often result in a high terminal voltage in the shack (equipment 'hot') on transmit and the mains supply acting as an earth return.

Test results: Compared with the station full-size inverted-vee dipole for 80 m and short inverted-vee dipole for 80m with 40m traps, the noise level increased as follows:

80m:	2.5 to 4 S-points
40m:	1 to 3 S-points

Antenna with a single-wire feeder 2:
Windom (at 36ft with single-wire down-lead to shack)

Although Windom and other antennas with a single-wire feeder may radiate well, the problems on transmit and receive are similar to those involved with standard end-fed antennas.

Test results: Compared with the station full-size inverted-vee dipole for 80m and short inverted-vee dipole for 80m with 40m traps, the noise level increased as follows:

80m:	2.5 to 4 S-points
40m:	1 to 3 S-points

Horizontal dipole for 80m

Height 36ft with 75 ohm twin feeder to a balun near ground level and 75 ohm coaxial cable to the shack.

Test results: Compared with the station full-size inverted-vee dipole for 80m, the noise level increased as follows:

80m:	1.5 to 3 S-points

Corrective actions taken: Neither of the following measures improved the received noise level: Insertion of a 1:1 balun at the feed point; use of 75 ohm coaxial cable to the shack instead of 75 ohm twin feeder.

Conclusions: This antenna has better noise immunity than an end-fed antenna or an antenna with a single-wire feeder, but cannot match an inverted-vee dipole, since the ends cannot be positioned for minimum noise pick-up.

Horizontal G5RV antenna

Height 36ft with 34ft length of 450 ohm ladder-line to feed point near ground level and 75 ohm coaxial cable to the shack (see Note 1).

Test results: Compared with the station full-size inverted-vee dipole for 80m and the short inverted-vee dipole for 80m with 40m traps, the noise level increased as follows:

    80m: 1 to 3 S-points
    40m: 0.5 to 1.5 S-points

Corrective actions taken: Each of the following measures improved the received noise level by 0.5 S-points:

- Reversal of feed-point connection so that the house end of the antenna was connected to the braid of the coaxial cable at the feed point.
- Inclusion of ferrite sleeve on the coaxial cable at the feed point, and reversal of feed-point connection so that the house end of the antenna was connected to the inner conductor of the coaxial cable at the feed point.

None of the following measures improved the received noise level:

- Earthing either side of the ladder-line at the feed point.
- Earthing the braid of the coaxial cable at the feed point.
- Positioning the ladder-line at an angle of 45° to the ground, parallel to the ground and vertically.

Conclusions: This antenna has better noise immunity than an end-fed antenna or an antenna with a single-wire feeder, but cannot match an inverted-vee dipole, since the ends cannot be positioned for minimum noise pick-up.

Inverted G5RV antenna

Height 36ft with 34ft length of 450 ohm ladder-line to the feed point near ground level and 75 ohm coaxial cable to the shack (see Note 1).

Test results: Compared with the station full-size inverted-vee dipole for 80m and the short inverted-vee dipole for 80m with 40m traps, the noise level increased as follows:

    80m: 0.5 to 2 S-points
    40m: 0.5 to 1.5 S-points

Conclusions: This antenna has a high degree of noise immunity because the ends can be positioned for minimum noise pick-up.

Vertical antenna

Height 60ft with series capacitor at base and 50 ohm coaxial cable to the shack.

Test results: Compared with the station full-size inverted-vee dipole for 80 metres and the short inverted-vee dipole for 80m with 40m traps, the noise level increased as follows:

    80m: 1 S-point
    40m: 1 S-point

Conclusions: Vertical antennas are generally considered to be noisy, probably because of their low angle of radiation and reception. However, this one was positioned in a noise-free zone well away from the house and other noise carriers.

Note 1: A balun is not essential on a G5RV antenna, but should be included to ensure that the length of the coaxial cable is not critical on the lower frequency bands.

Quarter-wave inverted-L for 160m

Length 170ft with series capacitor at base and 75 ohm coaxial cable to the shack.

Test results: The noise level across the 160m band was S4-5.

Conclusions: This noise level was considered to be in line with the levels achieved on this band using a random length long wire, and on the 80 metre band using other antennas.

Summary of test results

Antenna under test	Increase in S-points				
	160m	80m	40m	30m	20-10m
Random-length end-fed wire	*	2.5 - 4	1 - 3	**	***
Windom		2.5 - 4	1 - 3		
Horizontal 80 m dipole		1.5 - 3			
Horizontal G5RV		1 - 3	0.5 - 1.5		
Inverted G5RV		0.5 - 2	0.5 - 1.5		
Vertical		1	1		

*   The station 50-75 ohm Z-match would not provide a match on this band, so another external Z-match was used. The actual noise level (S 5-6) would almost certainly be bettered by a centre-fed antenna.

**   The actual noise level (S 1.5) would hardly be bettered by a centre-fed antenna.

*** The actual noise level (S 0.5) would hardly be bettered by a centre-fed antenna.

Overall Conclusions

Frequencies most affected by noise

The above-mentioned tests clearly confirmed that the electrical noise level tends to be far higher on the low-frequency bands, ie below 10MHz.

Factors not affecting noise immunity

Where the horizontal G5RV antenna was concerned, it was felt, initially, that the degree of feeder imbalance or mere verticality of the 34ft ladder-line could be responsible for the higher noise level. This was disproved.

Antenna Deficiencies

Antennas with a single-wire feeder

Even though the down-lead (the antenna itself) was brought into the shack (a shed) in a noise-free zone, the noise level on receive was high. It would have been a lot higher if the down-lead had been brought into a noise zone or in close proximity to noise carriers.

Therefore, regardless of where the shack is located, end-fed antennas and antennas with a single-wire feeder are not to be recommended.

Centre-fed horizontal antennas

These antennas have greater noise immunity than an end-fed antenna or an antenna with a single-wire feeder, but cannot match an inverted-vee dipole or inverted G5RV antenna, since the ends cannot be positioned for minimum noise pick-up.

Even though the S-meter readings for the station antennas and the antennas under test were comparable on certain frequencies, the *audible* noise level always seemed markedly higher with the antennas under test.

Preferred Antennas

Centre-fed antennas with an inverted-vee configuration

(Note: Voltage points should, in any case, be kept out of reach for safety reasons.)

- **Flexibility with regard to position:** All inverted-vee centre-fed antennas are very flexible from the positional point of view. That is to say:
  - the feed point can be positioned away from the house
  - the antenna can be 'rotated' around the feed point
  - the antenna can take the form of a straight line, a vee or a half-swastika
  - the ends can be positioned for minimum noise pick-up.
- **Flexibility with regard to frequency:** Since an inverted G5RV antenna is more flexible than an inverted dipole with regard to frequency, it was decided to carry out further noise immunity tests on the G5RV.

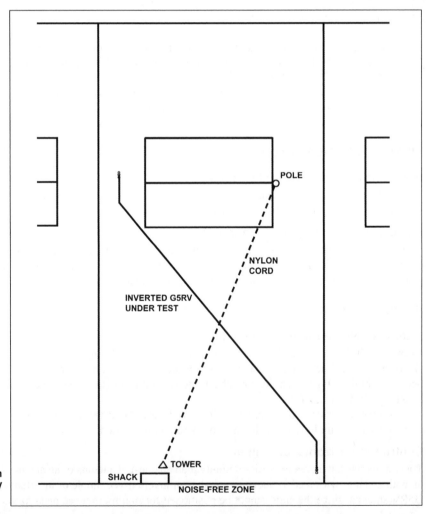

Initial configuration of inverted G5RV under test

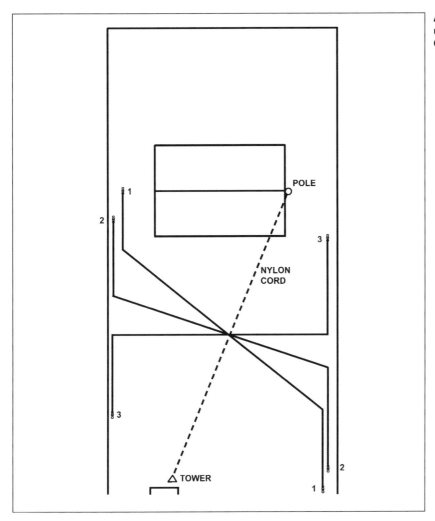

Alternative config-urations of inverted G5RV under test

Further Antenna Tests and Final Conclusions

All the tests specified in this section were carried out on the 80m band.

Test set-up using an inverted G5RV antenna

As stated above, all the antennas involved in the initial tests, apart from the vertical antenna, ran from a pole on the side of the house to a point on the tower near the shack at the bottom of the garden.

For the following tests, however, a thick nylon cord ran from the pole to the tower, and the feed point of the inverted G5RV antenna was suspended at various points along the cord. One end of the antenna was secured to a point in the far corner of the garden away from the house, and the other end to a point near the rear wall of the house, as shown opposite. Several other antenna configurations are possible, as shown above.

Noise measurements using portable wide-band noise source

A wide-band noise source, consisting of an old battery-powered vibrating door bell, was used for the measurements. It is pictured below with its antenna.

**Portable wide-band
noise source**

Test 1:
Effect of spacing between stand-alone noise source and feeder / ends of antenna

The noise source, with 1m long whip antenna, was positioned at various distances from the 450 ohm ladder-line. It produced the following noise levels in the IC-756 Pro3 receiver:

    1m = S5
    2m = S4
    3m = S1.5

The noise source, with 1m long whip antenna, was then positioned at various distances from the ends of the antenna. It produced the following noise levels in the IC-756 Pro3 receiver:

    1m = <S9
    2m = >S5
    3m = <S4

It was thus concluded that the feeder of an antenna has greater noise immunity than the ends, presumably because it is inherently balanced.

Effect of connecting noise source to live, neutral and earth leads of mains wiring

The noise source, without the whip antenna, was then connected in turn to the live, neutral and earth wires of the domestic mains supply, thus causing the entire house to become a virulent noise carrier. Since the received noise level was very similar whichever wire was connected, it was decided, for reasons of safety, to inject noise into the earth lead alone for Test 2.

Effect of connecting noise source to different power outlets in the house

The noise source, without the whip antenna, was then connected to various power outlets on the upper and lower floors of the house, and it was noted that the received noise level varied by up to four S-points. The differences were almost certainly caused by the configuration of the house wiring. For reasons of uniformity, only one point could be used for Test 2, and the one nearest to the rear door of the house was chosen since it was the most convenient.

Test 2:
Effect of spacing between rear wall of house and feeder / near end of antenna

With the noise source connected to the earth lead of the above-mentioned power outlet, the noise level was measured with the feeder and near end of the antenna at various distances from the rear wall of the house. The results are shown in the following table.

| Feeder | Near end of antenna | | |
	Touching	6m	10m
1.5m	S8	<S6	S5
3m	>S7	<S6	<S4
4.5m	>S7	>S5	<S4
6m	>S7	S4	<S4

Final Conclusions Regarding Antenna Noise Immunity

The above-mentioned measurements clearly showed that:

- the feeder of an antenna has greater noise immunity than the ends, almost certainly because it is inherently balanced.
- even virulent noise sources are relatively innocuous unless they are connected to house wiring, domestic antenna systems, overhead power-distribution cables or overhead telephone lines, all of which then become noise carriers.

Every effort must therefore be made to maximise the spacing between the feeder / ends of an antenna and the nearest noise carrier.

Signal Levels Achieved Using Various Configurations of the Inverted G5RV Antenna

The signal levels achieved using the inverted G5RV antenna at 36ft, regardless of its configuration, compared well with those obtained using the station full-size 80m inverted-vee dipole at 40ft and the short inverted-vee dipole for 80m (with 40m traps) at 38ft. They were often on a par with the levels achieved using the station wire antennas and rarely more than two S-points down.

6

Electrical noise from outside the premises

Radiation and Conduction of Electrical Noise

Almost all the output power developed by a radio transmitter is radiated as a signal, and only an infinitesimal amount, if any, is conducted through the power distribution system. This is because the transmitter is designed solely for radiation.

The electrical noise generated by faulty or inadequately suppressed electrical devices, for example motors, ignition systems, engine-tuning systems and heating and power control and generating systems, may have a relatively high energy level at source and may be radiated locally. However, unlike radio signals, *it may also be conducted over power-distribution cables or other wired systems*.

It has been found in practice that most electrical noise above 10MHz is radiated, and most electrical noise below 10MHz is conducted.

Reducing Noise Conducted to an Amateur Station via the Mains Supply

Use of filters at the mains input to premises and / or shack

It is always advisable to install a heavy-duty filter at the mains input to your premises and / or a smaller filter at the mains input to your shack.

The mains earth, provided by an electricity supply company at the point where the mains supply enters a property, may conduct electrical noise from nearby premises and thus increase the noise level on receive. It must therefore be filtered as well as the live and neutral leads. The following filters meet this requirement, *but have exposed terminals and should therefore be suitably enclosed.*

Mains input to your property: RS Components type RS 239-775, rated at 60A. (Most modern domestic systems would require a filter rated at 80 or 100A.)

A mains filter obtainable from RS Components (uk.rs-online.com)

> *Warning: Anything installed at the mains point of entry should be approved by the local electricity authority, and the work should be carried out by a competent registered electrician.*

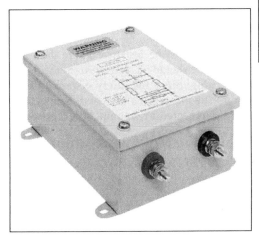

Belling-Lee mains filter obtainable from Farnell (uk.farnell.com) and other suppliers

Mains input to your shack: Belling-Lee type SF4240-16/01, rated at 16A. (*Note: This filter is designed for inclusion in equipment, and should therefore be enclosed. If installed on a ring-main, it should be protected by a suitably rated fuse.*)

Good shack-keeping

Components and wiring must never be allowed to form a 'birds nest' in the shack, and the number of leads must be kept to a minimum.

Use of capacitors across mains input to neighbours' premises

It is good practice to install suitably rated capacitors across the mains input to your neighbours' premises (as described and illustrated in Chapter 4), provided that they agree.

Warning:

Care should be exercised to ensure that RCD devices are not tripped through leakage. Working on high voltage electricity supplies can be dangerous and should only be done by someone suitably qualified. Care must be taken to comply with the regulations relating to the mains power supply in your country.

Checking for Noise Conduction Via the Mains Earth

The mains earth may be disconnected very briefly so that the received noise level can be checked while it is connected and disconnected. If the noise level improves with the mains earth disconnected, an alternative earth can be installed by driving a 1m long copper earth rod into the soil at a suitable point, and connecting it to the earthing system of the house via a thick insulated copper lead. This earth must be installed before the original earth is permanently disconnected, and should be separate from the one used for the station equipment.

Warning: Since the safety system of domestic properties is very dependent on the integrity of the earth connection, it should only be disconnected and reconnected by a competent registered electrician.

Protective Multiple-earthing (PME) Systems

The earth of PME systems is effectively in parallel with the neutral, and thus shares the normal neutral current. For this reason alone, the above-mentioned guidelines do not apply. Anything installed at the mains point of entry should be approved by the local electricity supply authority, and the work should be carried out by a competent registered electrician.

Natural Attenuation of Conducted Noise

Noise carriers such as power-distribution and other wired systems will almost certainly have very poor characteristics at radio frequencies, and may run for considerable distances underground or through buildings.

Conducted noise will therefore be subject to a high degree of attenuation which will obviously increase with frequency. In fact, although electrical noise may be conducted over a distance of several kilometres at 1MHz, and up to 1km or more at 3.5MHz, it may only be conducted a few hundred metres at 28MHz.

The residual noise will manifest itself at points where the above-mentioned noise carriers run above ground, eg sub-stations, overhead power-distribution cables and overhead telephone lines, etc, the so-called 'hot spots'. Long-term, first-hand experience has shown that these points are rarely the actual noise source (see Chapter 2).

Radiation from Noise Carriers

Because of the above-mentioned attenuation, the residual noise energy at the noise carriers will be considerably lower than the level obtaining at the noise source itself, and will therefore be radiated over a much shorter distance, perhaps only some 30 metres.

Although this residual noise can easily be detected at close range using a sensitive short-wave receiver with a whip antenna, it will usually provide no indication whatsoever of the actual noise source.

Analysing Electrical Noise

Although the type of electrical noise, its duration and the time of day at which it occurs can provide an amateur with some idea of its origin, he should never spend a lot of time trying to analyse it. After all, even if he is 100% certain that it originates from an unsuppressed motor, for example, he will still have to find the motor!

Initial Checks to Locate a Noise Source

If broad-band or any other intrusive type of noise is detected, one's own property should be checked to ensure that it does not contain the noise source. This can be done by running the transceiver, or a separate receiver, from a 12V accumulator battery while the mains supply to the property is switched off, thus isolating it. Obviously, modern transceivers using a separate 13.8V supply make this task a lot easier.

If one's own property is found to be in order, each of the adjacent properties should then be isolated from the mains supply in turn to ensure that they do not contain the noise source. Most neighbours are only too happy to participate in this procedure which is simple and which can usually be carried out at a time which causes a minimum of domestic disruption.

If a property is found to contain the noise source, each piece of equipment in that house must then be disconnected off *in turn* until the offending item is found. The appropriate action can then be taken to eliminate the noise, for instance by applying filters, ferrite rings and clip-on ferrites (as described Chapter 9).

Direction-finding to Ascertain the Approximate Location of a Noise Source

If local electrical noise remains a problem, the next task is to ascertain the approximate location of the noise source. This can only be done by direction-finding. Details of the direction-finding procedure, and ways of improving the accuracy of measurement, are shown in Chapter 7. The basic procedure is as follows.

A loop antenna must be set up in an open space, a large garden or park for example, and a bearing taken on the noise. It is inadvisable to take a bearing in the close vicinity of noise carriers, since they could adversely affect the result.

Once the direction of the noise has been ascertained, a similar open space must be selected at right-angles to the line, and another bearing should be taken to obtain a fix. This will be the *approximate location* of the actual noise source.

Tracking Based on the Amplitude of the Noise

This method is very unsatisfactory, since the amplitude of the noise will tend to increase in the proximity of noise carriers, eg electrical sub-stations, overhead power-distribution cables, overhead telephone lines, etc, and the person involved in the tracking operation will have no means of knowing whether he is walking towards the noise source or away from it. Why waste shoe leather walking the streets looking for noise carriers?

Tracking Based on the Maximum Frequency of the Noise

Tracking based on the maximum frequency of the noise is described in the *Interference Handbook* by William R Nelson, WA6FQG (Editor, William I Orr, W6SAI). This handbook recommends that a small portable receiver be used for tracking, the smaller the better so as not to draw the attention of residents. It may have analogue or digital tuning, must be sensitive, and should preferably tune to 30MHz or higher. The AM or SSB tuning mode may be used.

The handbook suggests the following distances from a noise source at which electrical noise can be heard: 3.5MHz - 1100m, 7MHz - 800m, 14MHz - 580m, 28MHz - 330m and 144MHz - < 150m.

When direction-finding has been used to pin-point the *approximate location*, the use of a small sensitive all-band receiver and whip antenna may well reveal noise at frequencies higher than the original noise frequency. If so, it will merely be necessary to increase the frequency while walking until a point is reached where the noise is at its highest frequency. This will be the *actual noise source*!

Local Plot Plans

Local plot plans are invaluable for plotting the direction of noise sources. They can usually be obtained very cheaply from a Local Authority or Borough Council.

7

Direction-finding equipment

Loop Antennas in General

Many loop antennas are unsuitable for direction-finding because they are connected directly to coaxial cable which is inherently unbalanced. This causes distortion of the directional pattern, and meaningful nulls cannot be obtained.

The Wellbrook loop antennas (see www.wellbrook.uk.com) are different in that the screen of the coaxial cable used to connect the loop itself to the interface unit in the shack is isolated. This ensures that there is no distortion of the directional pattern.

Description of Wellbrook Loop Antennas

The ALA1530 and ALA330S antennas are approximately 1m in diameter. They are light, robust and completely waterproof. They can therefore be used outdoors. They need no tuning whatsoever, can be positioned well away from the house and other noise carriers, and can easily be rotated for direction-finding purposes, and for nulling out co-channel QRM, adjacent-channel QRM or local electrical noise. Both loops can be powered from a 12V car battery for portable direction-finding purposes.

Sensitivity

Both loops are extremely sensitive. Indeed, if the plane of either loop is in line with a received station, it is often impossible to tell whether it is being received on the loop or on the main antenna.

Directivity

Although the gain is high in the direction of the loops, the S-meter reading is much lower when they are at right-angles to a received station. The forward and backward lobes of both loops are fairly broad and the side nulls are sharp, so they are ideal for direction-finding purposes and as a back-up on receive.

Model ALA1530

The advertised response of the ALA1530 is substantially flat from 100kHz to 30MHz, and stations can be received at good strength from lower than 50kHz to well over 30MHz. It permits good reception of the Loran navigation system on 100kHz as well as the whole of the Long, Medium and Short Wave bands. However, the gain tends to drop off slightly at the higher frequencies, ie above 10MHz. The loop is ideal for general listening and monitoring purposes

Model ALA330S

The ALA330S is a more recent version of the loop which provides greater gain than the ALA1530 from 3.5 to 30MHz, in fact a little too much gain at

3.5MHz! It provides similar gain to the ALA1530 at 1.9MHz, but much less gain below 1.6MHz.

If the ALA330S loop is used, it is advisable to connect a 100 ohm carbon potentiometer across the output of the interface unit, with the slider connected to the receive port of the transceiver. This will permit the loop gain to be adjusted. The potentiometer should be set to mid-position for 80m, to mid-position or higher for the other bands and to near-zero or zero for direction-finding. This loop is ideal for amateur radio purposes.

Accessories supplied

The Wellbrook loop antennas are supplied complete with an interface unit, a stabilized 12V power adapter which is noise-free, and a short coaxial lead connecting the interface unit to the receive port of a transceiver.

The only additional item which has to be purchased is the RG58 coaxial patch lead (with BNC connectors) between the head amplifier on the loop and the interface unit in the shack. This item is not supplied because the distance between the loop and the shack will vary in accordance with the size and layout of the property involved. A 10m long coaxial patch lead with BNC connectors (available from Maplin) is ideal, but leads up to 30m long may be used.

Self-amalgamating tape may be used to seal the BNC connection at the loop end of the patch lead, thus preventing the ingress of water and dirt. If carefully applied, the seal will remain intact over a period of several years.

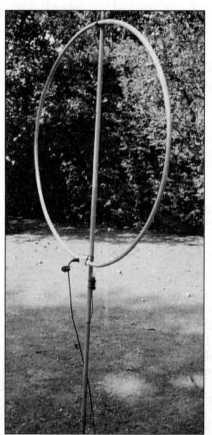

A Wellbrook loop antenna under manual control

Installation of Wellbrook Loop Antennas

The ALA1530 and ALA330S loops may be rotated manually or, more conveniently, using an in-line rotator such as the Yaesu G-250 which is ideal for the purpose.

Manual rotation

If the loops are rotated manually, they may be installed on a wooden pole near ground level. The photograph shows an ALA1530 loop mounted on a length of wooden dowelling, 2cm in diameter, which is coated with two layers of clear varnish. Four large black plastic ties are used to secure the loop to the support, two at the bottom and two at the top. The bottom of the loop is about 1m above the ground.

As shown in the photo, a jack plug is mounted on the support just below the loop. This permits a pair of headphones to be connected via twin lead to the transceiver headphone jack. Headphones may thus be used to monitor the noise level while the loop is rotated manually for a null.

Remote-controlled rotation

If the loops are turned using a rotator, the aluminium pole supporting the rotator should be between 1³/₈" and 1¹/₂" in diameter and approximately 9 to 12ft high. The base of the pole should be embedded in concrete. A wooden pole approximately 1¹/₈" in diameter, and coated with two layers of clear varnish, should be mounted on top of the rotator. The loop should be attached to this support as described above.

When a rotator is used, the position of the loop must be carefully set by taking a null on a local medium-wave broadcast station whose exact location is known. This information can be obtained from the *World Radio and Television Handbook* (www.wrth.com).

Use of Wellbrook Loop Antennas

The directional pattern of the Wellbrook loop antennas, models ALA 1530 and ALA 330S, is a true "figure-of-eight", with sharp nulls at right-angles to the plane of the loops, which can thus be used for the following purposes:

A Wellbrook loop antenna under remote control

Direction-finding

The loops can be used for direction-finding in order to trace:
  - local sources of broad-band noise, narrow-band noise on random discrete frequencies, and narrow-band noise on groups of frequencies.
  - local or remote sources of deliberate or malicious interference.

When the loops are used for direction-finding, it is good practice to take a null, and then add or subtract 90 degrees, since the nulls are sharp, whereas the forward and backward lobes are fairly broad.

In the case of broad-band noise, it is advisable to carry out direction-finding at the lowest frequency at which the noise is heard, thus ensuring greater accuracy. It is often easier to take a null in the AM mode (that is to say, by reducing the level of AC hum) than in the SSB mode.

Back-up on receive

The loops can be used as an alternative antenna on receive to permit contacts to be maintained when reception via the main antenna is difficult or impossible because of band conditions, ie co-channel and / or adjacent-channel interference and / or local electrical noise.

Spectrum monitoring

The loops can be used for monitoring the spectrum by switching a transceiver in 500kHz steps, say from 500kHz to 60MHz, and noting the wide-band noise level on the S-meter at the various 500kHz points. Some modern transceivers have up to 100 memories, and the required frequencies can thus be stored to facilitate this operation. The spectrum monitoring transceiver should preferably be set to the SSB or AM mode for this purpose.

Limitations on the Use of Wellbrook Loop Antennas

The following paragraphs show limitations on the use of Wellbrook loop antennas and should be carefully noted.

Eliminating possibility of connection to RF sources

The loops must be connected to the receive side of the transmit-receive relay on older transceivers, or to a dedicated receive port on modern transceivers. Even a microwatt of RF in the cable connecting the loop will destroy the head amplifier instantly!

Eliminating RF currents on RG58 patch lead

For protection purposes, a pair of ferrite rings or a number of clip-on ferrites should be installed at each end of the RG58 patch lead between the loop itself and the interface unit, thus isolating any RF currents which may be present on the outer braiding during transmissions.

Eliminating interaction with other antennas

If any interaction is noted between the loops and adjacent unused antennas during direction-finding operations, the unused antennas must be shorted out at their feed point.

Eliminating rectification due to strong RF fields

Strong RF fields in the vicinity of the loops may cause rectification to take place in the interface unit, thus producing a small DC voltage across the coaxial lead between the interface and the receive port of the transceiver. This could block reception. However, the problem is easily solved by including a pair of 1N4001 diodes in a back-to-back (antiparallel) configuration across the output of the interface unit.

Conclusions Regarding Wellbrook Loop Antennas

The loops will generally be used:
- as a back-up antenna for nulling out co-channel and adjacent-channel QRM when reception becomes difficult on the main antenna, or
- to null out local noise which would otherwise make it difficult or impossible to maintain contacts.

They will sometimes be used:
- to indicate the direction of a local noise carrier or source so that a noise antenna can be pointed in the right direction (see Chapter 8);
- to indicate the direction of a source of deliberate or malicious interference; or
- for spectrum monitoring.

8

Noise-cancelling equipment

Noise-reducing Devices (DSP)

Noise reduction (DSP) does not eliminate noise, but reduces noise components and picks out wanted signals which are buried in noise. In other words, it merely alleviates problems caused by the noise, since the noise is still there.

As with computer viruses, prevention is a far better cure, and a well-positioned centre-fed antenna, used in conjunction with a directional loop and noise canceller, will provide a much better solution in the long run, since it really does enable noise to be eliminated.

Description of Noise Cancellers

These devices permit local noise picked up by a (main) station antenna to be compared with the same noise picked up by an auxiliary noise antenna (or antennas). The noise inputs can then be balanced so that unwanted interference is partially or completely cancelled.

Noise cancellers are excellent for reducing or removing the following types of local noise:

- Broad-band noise, including PLT
- Narrow-band noise on given discrete frequencies, and
- Narrow-band noise on groups of frequencies.

There is no point in purchasing a noise canceller with pre-amplifier, since the pre-amp will merely cause intermodulation and degrade the front-end performance of the receiver.

Limitations of Noise Cancellers

A noise canceller will only:

- reduce or remove local broad-band or narrow-band noise *from one source at a time*. Thus, if it is tuned to remove local noise on a given frequency,

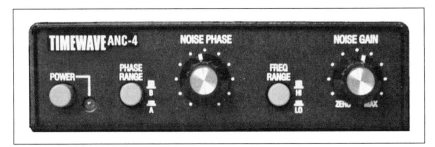

A commercially-available noise canceller, the Timewave ANC-4

Another commercially-available noise canceller, the MFJ-1026

3696kHz for example, it will need to be retuned to eliminate local noise from other sources in other parts of the band.

- provide optimum noise elimination if the level of noise picked up by its auxiliary noise antenna or antennas is at least equal to the level picked up by the main (station) antenna in use.

Practical Auxiliary Noise Antennas

A loop antenna may be used to ascertain the direction of each main noise source so that a noise antenna can be pointed towards it. The noise picked up by the station antenna and the appropriate noise antenna can then be balanced and nulled out.

As a more flexible alternative, the author has erected a permanent array of three auxiliary noise antennas spaced at an angle of 120 degrees to each other, thus providing full coverage over 360 degrees.

The auxiliary noise antennas each run from the shack to a point 25ft high on the tower before sloping down in different directions. The bottom end of each noise antenna is fixed to the top of a copper earth rod, the bottom half of which is embedded in the soil.

Auxiliary noise antenna selection and matching

A typical noise antenna selector, consisting of two rotary switches in a plastic box, is shown in Chapter 4. It can be used to select any one of three noise antennas, or any two of them in parallel. Output from the noise antenna selector is connected to the noise canceller via a small matching unit (see photographs opposite) which can be used to tune the selected noise antenna (or antennas) to achieve optimum noise cancellation.

Two views of the noise canceller matching unit

Local electrical noise is generally at its highest level on 160 m, slightly lower on 80 m and lower still on 40 m. The arrangement shown permits more than 90% of local electrical noise to be eliminated.

Problems Due to Inadequate Spacing Between Main (Station) and Noise Antennas

It is advisable not to install the auxiliary noise antennas too close to the main (station) antennas, since the resulting high induced voltage could well cause the protective lamp, in series with the noise canceller input circuit, to burn out. A spacing of some 15 to 20 feet would seem to be adequate.

Because of the above-mentioned limitation, a noise canceller cannot be used with end-fed and Windom antennas, since the down-lead and single-wire feeder of those antennas and the auxiliary noise antenna(s) of the canceller have to be brought into close proximity in the shack. It is then impossible to transmit without blowing the protective lamp in series with the noise antenna input circuit of the canceller.

The protective lamp should, in any case, be mounted in a small two-pin socket on top of the noise canceller. It can then be monitored during balancing operations to minimise the possibility of burn-out and permit easy replacement.

Other Uses

The auxiliary noise antenna and main (station) antenna gain controls of a noise canceller can be used not only for balancing and cancelling purposes, but to permit a direct comparison to be made between the electrical noise levels produced by the low-impedance station antennas and the high-impedance end-fed noise antennas. The levels may differ by a couple of S-points or much more, depending on local conditions.

9

Interference to domestic equipment

This chapter may well appear to be out of line with the main purpose of this book, ie noise elimination. However, it must be remembered that electronic devices which are susceptible to interference from amateur transmissions are very often the ones which generate electrical noise affecting reception. For this reason alone, it is always advisable to fit filters and / or ferrites to a neighbour's equipment *free of charge*.

Precautions to be Taken in the Shack

The following points should be borne in mind before protective devices are applied to domestic equipment.

Connections and earthing

A transmitting station must always be carefully interconnected and earthed to ensure that all enclosures, including the braiding of interconnecting cables, are at earth potential. Components and wiring must never be allowed to form a 'birds nest' behind the station equipment, and the number of leads must be kept to a minimum.

Low-pass filters

In the days of Band I television (45MHz vision and 41MHz sound), operation on 14 and 21MHz was virtually impossible during television hours in south-east England. This was because amateurs had great difficulty in suppressing the second and third harmonics when operating on those bands.

Even today, transceivers are rarely designed to prevent harmonic leakage, so low-pass filters fail to work because harmonics of the transmitted signal on the outside of the enclosure are able to by-pass the filter on their way to the antenna. In all my years of amateur radio operation, I have never come across a low-pass filter that worked.

Protection Applied to Domestic Equipment

Filters and ferrites

Interference may be received via a domestic antenna system or via the mains supply. The following protective devices should therefore be applied to your own and your neighbours' domestic equipment in order to eliminate interference during transmissions.

- **Antenna down-leads:** High-pass filters and/or clip-on ferrites should be installed at the bottom of antenna down-leads.
- **Interconnecting leads:** Clip-on ferrites should be installed on interconnecting leads.

- **Mains power leads:** Ferrite rings should be installed at the point where power leads enter equipment.

Other Means of Protection

Earthing

It may be necessary to earth domestic equipment, in which case a standard copper earth rod should be used. The earth lead should preferably be as short as possible and connected to the equipment at only one point.

Changing the position of affected equipment

- **Touch lamps:** These produce emissions similar to switch-mode power supplies, but with a fundamental frequency around 190kHz. Although they do not cause noise problems as such, they are easily switched on and off by radio transmissions on various frequencies from antennas located at distances up to 30m or more. Ferrite rings and clip-on ferrites will rarely effect a cure. The problem can be overcome by moving affected lamps away from large metallic objects such as bed frames, other power leads, power outlets, water pipes, window frames, etc.
- **Alarm systems:** These do not cause noise problems as such, but may be triggered by radio transmissions on various frequencies from antennas located at distances up to 30m or more. The problem can be overcome by moving affected sensors away from large metallic objects such as power leads, power outlets, water pipes, window frames etc.

Screening (not recommended)

If transmissions interfere with your own or your neighbour's audio or video systems, no attempt should be made to resolve the problem by screening the leads (to speakers for example). This will only increase the level of interference due to common mode signals on the braiding.

Telephone carrier systems (not recommended)

If a second telephone line is required, it is often possible to use a spare pair of conductors in the existing telephone cable for the purpose. However, if there are no spare conductors, BT will sometimes suggest the use of a carrier-derived circuit to avoid the need for a new cable.

Unfortunately, telephone carrier systems are very susceptible to radio frequency interference, and the installation of a new multi-pair cable is generally essential.

Protection of domestic systems resonant in or near an amateur band

Domestic television or hi-fi systems with an indoor (attic) or outdoor antenna may be resonant in or near an amateur band. If the high level of noise emanating from a system or its extreme susceptibility to interference suggests resonance, it can be verified by applying hand capacity to the antenna down-lead by gripping the coaxial cable tightly during transmissions.

If resonance is suspected, high-pass filters, clip-on ferrites and / or ferrite rings should then be installed at crucial points on the system in order to eliminate it.

10

Noise from Computers

My own computer equipment is, for historical reasons, located some distance from my main shack. To reduce potential interference to a receiver close to the computer, I carried out the measures described earlier in this book.

Ferrite rings were fitted to the mains leads of the computer, associated printer and scanner. All interconnecting leads to the keyboard, mouse, printer and screen were fitted with clip-on ferrites. The system has worked well for many years, and I have no desire to move it into the main shack.

However, many amateurs use computers in their radio shacks nowadays. Typical uses are:

- As a logging system, to record details of all previous and subsequent contacts for future reference.
- To provide access to the web, for instance for *qrz.com* or *eHam.net*.
- Control of the radio and rotators etc.
- Software-defined radio.

My old friend Reg Stevens, G3TDH, worked as an EMC engineer for several years, and uses a large room in his house for both amateur radio and computer equipment. As a result of a conversation we had, Reg very kindly wrote the following notes covering possible EMC problems which computers could cause for radio amateurs.

Modern shacks may have one or more computers which may give rise to noise problems

Computers and EMC

Until the 80s, amateur radio had very little association with the digital computer world. However, the technology has become ubiquitous and almost every household in the land seems to have at least one machine in use, and often more than one. The same can be said of amateur radio stations where a PC has become a normal addition to the plethora of electronics that most amateurs take for granted. Indeed, the machines have moved from being a more efficient alternative to the old paper logbook, have popularised digital transmission and, since the advent of the SDR, have become the basis of the station itself.

Like most electronic equipment, computers emit and are susceptible to RF interference, and EMC is a significant issue. Nevertheless, most properly designed and installed machines contribute no more noise to the environment than anything else. The biggest issue, as in so many other cases, is their dependence on switch-mode power supplies. Most desk-top machines have a well-shielded internal power supply, whereas the lap-top equivalent usually employs an external unit which is either housed in a rectangular plastic box in the power lead, or a supply incorporated in an over-sized mains power outlet, commonly referred to as a "wall wart".

In general, machines sold under well known brand names are well designed and meet the supposedly mandatory EMC standards in a proper manner. If they are installed in close proximity to the shack antenna or feeder system - a likely scenario if constituting part of an amateur radio station - additional filtering and careful attention to layout is essential. Not only can a computer produce background noise and rough unstable 'birdies' over amateur frequencies, but may become unstable or crash in the presence of a strong RF field. However, the most common noise problems are switch-mode power supplies and unshielded network cables. Although most PSUs cause no difficulties whatsoever, there are still plenty of 'rogue' units that can make radio reception impossible, particularly on the lower frequency bands!

The power supplies of some desk-top machines come nowhere near the required emission standards, even though they have a CE label. Many are perfectly well designed, but the manufacturer has left out the necessary filter

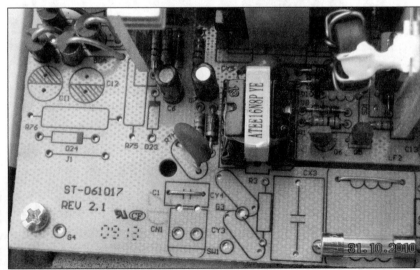

This rogue power supply has its board marked for the installation of EMC components but they have been omitted

components which usually consist of a double-wound choke on a ferrite core and two decoupling capacitors. When these components are present, the supplies cause no interference, but 'rogue' equipment can cause interference over a very wide area indeed. Modestly priced replacement power supplies are easy to find and easy to fit. If a machine causes problems, this option is much more effective than attempting to fit additional external filtering.

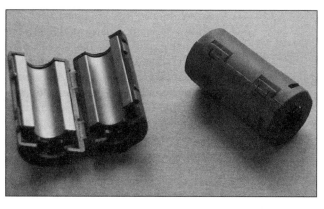

Clamp-on ferrites can be useful to reduce noise radiation (and pick-up) on computer leads

Lap-top machines suffer just as badly although, in my experience, the problem usually occurs when the power supply provided with the machine has to be replaced after becoming lost or broken. Cheap replacement power supplies obtained via *eBay* and the like are often poorly filtered. Again the simplest and most effective cure is to replace such supplies with a higher quality alternative, preferably one of a trusted and well-known brand.

It is less common for computers themselves to cause problems, but attention must be given to their peripherals and their power supplies as well. Faulty devices can easily be identified by switching them on and off, although the system containing these parts has to be tracked down first. Obviously, the first step is to make sure that nothing in your own household is causing problems before you start to search the district for troublesome neighbours!

Many home systems employ networks to connect the various units together. Such networks now extend to things more associated with television systems than computers, but frequently share the LAN provided for the computers of the household. Thus, smart TVs, media servers and digi-boxes supplied by Internet broadband providers are often networked together. When this is done using power-line transmission (PLT), there is obviously potential for tremendous local short-wave interference. However, even properly installed Internet cables of the RJ45 type can be problematic. Unlike PLT, these work with properly balanced differential signals, but can still cause problems over a short distance. They are best kept out of an amateur radio shack and, if used at all, should be kept as short as possible.

There is often a short RJ45 cable between ADSL modems and their associated router, which may be wi-fi or cabled. Although a short cable is not usually a major problem, longer lengths joining one part of a building to another may be difficult to filter on the lower frequency bands. Several turns on a high-permeability toroid will improve matters dramatically, but will be needed at both ends of the cable. Also, the toroidal rings will need to have a large centre hole to permit the relatively large plugs involved to be fed through. It is much better to replace such cables with a wireless (wi-fi) system. Although these cost a little more than the simple cable they replace, most lap-tops and later model TVs are already suitably equipped. Wi-fi is also a good deal tidier than either the cable or PLT alternatives and, if you become involved with a neighbour's system, the lady of the house is more likely to be your ally.

Computers need display screens, and some earlier CRT models can often be as problematic as TVs, with time-base oscillators which themselves serve as switch-mode PSUs providing EHT. Although modern LCD models do not generally cause too many problems, they still need checking, along with the inevitable plastic box that supplies their power. Again, it is worth mentioning that the addition of sufficient common-mode inductive reactance in the form of a ferrite clamp or ring can work wonders. However, it may be difficult to provide sufficient inductance to be effective on the lower frequency bands. In addition, it is not uncommon for the screens themselves to cause interference at VHF and no amount of clamps will help. Unfortunately, in such cases, replacing the screen with one with lower emission might be the easiest solution. Fortunately, most screens are of the common SVGA standard, and trying one from another system may prove the point.

11

Assistance and complaints

Sources of Assistance

The main source of assistance to radio amateurs suffering from electrical noise is the EMC website of the RSGB (http://rsgb.org/main/technical/emc/).

RSGB EMC website

The front page of the RSGB EMC website, includes the following useful links:
- How to track down interference
- How to avoid causing breakthrough
- Individual help from EMC Technical Advisors
- Regular EMC column in Radcom – 'EMC Matters'
- RSGB website self-diagnosis plus EMC leaflets

Obviously, you should do everything you can to track down the source and eliminate the noise before you call on an advisor. *Note that some RSGB EMC services are available only to members.*

Radio clubs

If you are fortunate to live in an area containing an active radio club, it may be worth attending one of the club evenings. Even if the members are unable to assist you directly, they may well be able to put you in touch with someone who can.

Complaints to Neighbours

If you suffer from a high level of electrical noise on receive, *and you are 100% sure of the noise source*, it is a good policy to broach the subject with the neighbour concerned.

The great majority of people are reasonable and, if you are equally reasonable with them, you will be able to come to a mutual agreement regarding the installation of filters and ferrites (free of charge), and the execution of suitable tests.

You should, in any case, ensure that your own domestic equipment is free of interference at all times when your station is operating at full legal power. If you then receive complaints of interference from a neighbour, you will be able to show him or her that your own domestic equipment is unaffected by your transmissions, but that you have to cope with severe noise deriving from his or her domestic equipment. Since the problem is reciprocal, it can be resolved to the benefit of both parties.

Simply avoiding the issue by reducing power is not a solution, nor is going off the air. It is much better to fix the problem promptly, at a mutually agreed time. If you fail to do so, the noise problem may never be resolved.

Complaints to Authorities and Organisations

Before any complaint is made, it is essential that you assist the authority or organisation concerned by ascertaining the approximate or actual location of the noise source using the direction-finding and tracking operations specified in Chapters 6 and 7. This will ensure that the matter is dealt with quicker and more efficiently.

Interference to broadcast TV or radio reception

Complaints are now dealt with by the BBC (see www.bbc.co.uk/reception).

Interference to amateur radio

Complaints regarding interference/noise should be submitted via the OFCOM website by filling in the form at:
- https://stakeholders.ofcom.org.uk/tell-us/abuse-amateur-radio-system

If OFCOM investigates but fails to eliminate a noise source for whatever reason, particularly one involving PLT or a plasma screen television, and the amateur licensee concerned is effectively prevented from receiving anything but very strong signals, there are two possible courses of action:
- The complainant can refer the case to the local electricity authority on the grounds that the mains supply is contaminated - see Chapter 4 and below; and / or
- He can arrange for articles to be published in local magazines or newspapers declaring the use of PLT to be antisocial in that it stops someone enjoying their legitimate licensed hobby and may well interfere with other users of the radio spectrum. The articles should preferably be submitted by someone other than a radio enthusiast! Note that this might be thought rather extreme, and therefore should only be considered carefully before action is taken.

Local electricity authority

As stated in Chapter 4, it is advisable to contact the local electricity authority if the mains supply is subject to long-term contamination by electrical noise from any source, whether intentionally as in the case of PLT, or otherwise.

BT and Virgin Media (cable systems)

The above-mentioned organisations are very cost-conscious when it comes to complaints of interference, especially from radio amateurs, and they will, in any case, have little experience of radio-frequency technology and EMC problems. The complainant must therefore be absolutely sure of the noise source before submitting a complaint regarding unwanted electrical noise.

12

Final recommendations

I have endeavoured to cover every aspect of electrical noise and every possible remedy. However, if asked to list briefly the most important factors to be taken into consideration when setting up a station, I would list the following without hesitation:

- ## Scenarios to be Avoided

  Avoid HV power lines and overhead power-distribution and telephone cables like the plague.

- ## Position of Station on Premises

  Always operate from a shed or caravan in a noise-free part of your premises.

- ## Noise Immunity of Antennas

  Always use centre-fed antennas, preferably with an inverted-vee configuration, and keep the ends, in particular, as far as possible from any noise carriers.

- ## Essential Ancillary Equipment

  Always use an all-band loop antenna and noise-canceller. If you are affected by electrical noise, a single transceiver plus an all-band loop antenna and noise canceller will always be far more effective in reducing electrical noise than any number of transceivers!

- ## Standing Up for Your Rights

  Let's face it, the ability to go on the air on any band, at any time of the day or night, using the maximum permitted power, without suffering from unwanted electrical noise on receive or causing interference on transmit, probably has far greater merit than all other aspects of amateur radio combined. Although most radio amateurs would love to go on the air at any time they choose, only a very small percentage of them actually do so. The reasons put forward for restricted operation are many and varied, but top of the list is almost certainly their unwillingness to confront their EMC problems and to stand up for their rights. Unfortunately, as stated previously, this attitude does nothing for our hobby.

Useful web links

Safety

- http://niceic.com/
  - The UK electrical contracting industry's independent voluntary body.
- http://www.esc.org.uk/
  - Electrical Safety Council.
- http://www.theiet.org/
  - Institute of Engineering and Technology.
- http://www.hse.gov.uk/electricity/index.htm
  - Health & Safety Executive.
- http://rsgb.org/main/files/2012/11/EMC07-final.pdf
  - 'Protective Multiple Earthing and the Radio Amateur' (RSGB).
- http://rsgb.org/main/files/2012/11/EMC12-Final.pdf
  - 'Part P and the Radio Amateur' (RSGB).
- https://stakeholders.ofcom.org.uk/tell-us/abuse-amateur-radio-system
  - Ofcom TV or radio interference or reception problems.

Noise

- http://www.radiolabs.com/Articles/amnoise.html
  - Radio Labs: 'AM or Shortwave Reception Problems? We may be able to help'.
- http://hfradio.org/ace-hf/ace-hf-noise.html
  - HFRadio.Org: 'Radio Signals, Noise and Effectiveness'.
- http://www.cv.nrao.edu/course/astr534/Discovery.html
  - 'The Discovery of Cosmic Radio Noise'.
- http://en.wikipedia.org/wiki/Noise_(radio)
  - Wikipedia on radio noise.
- http://www.radio-electronics.com/info/receivers/sensitivity/noise_floor.php
  - Radio-electronics.com: Tutorial on radio receiver noise floor.
- http://www.eetimes.com/document.asp?doc_id=1272302
  - *Electronic Engineering Times*: 'Understanding noise figures in receivers'.